An illustration from about 1760 showing the preparation of the hearth, three different chimney methods and the construction of the stack. Very similar illustrations appeared in French and English publications of this time, obviously copied. All ignore the covering of grass, straw or bracken which should precede the earth covering. Some operations are also out of order: the channel being dug around the kiln in the right foreground would have been made after the earth covering was complete and the kiln fired. The kiln at centre right does not appear to have a chimney.

CHARCOAL AND CHARCOAL BURNING

D. W. Kelley

Shire Publications Ltd

CONTENTS

Set in 9 point Times roman and printed in Great Britain by C. I. Thomas & Sons (Haverfordwest) Ltd, Press Buildings, Merlins Bridge, Haverfordwest, Dyfed.

British Library Cataloguing in Publication data available.

ACKNOWLEDGEMENTS
The author acknowledges the assistance of his many industrial contacts and friends for providing information and of his family for typing the text and, since he has no artistic talent, for preparing the sketches on pages 6, 14 (upper) and 26. Photographs on the following pages are acknowledged to: Shirley Aldred and Company Ltd, pages 26, 27; Simon Crouch and Border Television, cover; the Museum of English Rural Life, University of Reading, pages 9, 10, 11; the Weald and Downland Open Air Museum, pages 12, 13. All other illustrations are from the author's collection.

COVER: *A modern charcoal burn in the Lake District.*

BELOW: *A forest kiln site of between 1750 and 1850 near Dronfield, North Derbyshire. This site has been cut out of the hillside and levelled. Charcoal burners' pitsteads of this type are numerous in the woodlands of the area; four similar levelled hearths can be seen within 500 metres (550 yards) of this example. Charcoal made on this site may have been used in bloomeries or in the numerous forges which once lined the valley. The area is rich in ironstone and the charcoal may also have been taken by river to the Rotherham and Sheffield ironworks.*

An illustration of about 1760. The kiln in the left foreground is being fired. The kiln to its right is under way but the top has not yet been sealed. Behind this the contraction of the third kiln is shown while in the left background a fourth is near completion. All these are an artist's impression. The kilns were never as conical as shown here and the windbreaks are too low to be effective: kilns built in this shape would have been unstable.

ORIGINS

Charcoal is a black porous substance with a high carbon content made by heating wood or other organic materials, out of contact with air. The production of charcoal is one of the oldest crafts and the correct name of the process, the destructive distillation of wood, was arguably the first true chemical process.

The uses of charcoal in Europe can be traced back at least 5500 years. It was the smelting fuel of the bronze and iron ages. No other fuel then available to metalworkers could reach the temperatures needed, first to smelt the ore to make a *matte* and then to raise the temperature of the broken-up and washed matte to the point where the metal would melt and could be cast.

Implements and weapons made with iron were more durable and effective than those made from bronze. More efficient methods of producing iron brought improvements in armaments — all smelted, cast and worked by use of charcoal — so altering the course of civilisation. The short sword and body armour were instrumental in spreading the Roman empire and were early examples of large-scale production methods.

Roman writers recorded the uses of charcoal and the by-products of its manufacture. Pliny describes the use of wood pitch for caulking wooden ships. Vitruvius, a noted architect of the first century BC, recommended the charring of wooden piles and posts to preserve them from rot and water damage. He also recommended founding pillars and causeways, if they were to be built on soft or marshy ground, on footings of charcoal.

This knowledge came from the earlier Greek builders who based similar heavy structures on charcoal, many of which still stand today.

Zinc was smelted in medieval times in India, in furnaces heated to around 1100 Celsius, again using charcoal.

The liquor produced during the charcoal making process, pyroligneous acid, was used by the Egyptians and was an essential material in the embalming process. Temperatures greater than 1000 Celsius can be achieved easily with charcoal and it is reasonable to suppose that the Egyptians, who were expert metalworkers, would have discovered that sand mixed with other materials would melt at elevated temperatures to form glass. Since glass is made with sand, limestone and soda ash (sodium carbonate), which occur naturally in Egypt, it can be seen that charcoal was instrumental in the development of the early glass industry.

A kiln being prepared by the North African method. A shallow pit has been made and the base levelled. This stack is complete and will be covered with vegetation or straw before a covering of loose earth is spread over it. Note the neat stacking method and overall shape of this tapered stack. It will be ignited at the narrow end and burning progresses to the wider far end, taking ten to fifteen days to complete the process. The charcoal burner will keep the earth covering damp as the burning line moves down the kiln. If necessary he will cut down into the stack and divide off a completed section. Final quenching is effected by running water into the pit which results in a spectacular amount of steam but makes the charcoal dirty and wet. Since the product is sold by weight this is a distinct advantage to the burner, but the quality of the product is low.

A rectangular shallow pit kiln of the Nile delta, the wood covered with straw and ready for the earth covering. The second kiln to the rear is already partly covered. Behind this can be seen the remains of a completed burn after the charcoal has been removed. The burner did not live in the mud-brick building seen in the left background but used it to hold his tools and sacks. During the burn period he lived in the straw-roofed shanty on the right.

TRADITIONAL FOREST KILNS

The traditional earth-covered forest kiln is seen now in Britain only in demonstrations of the ancient craft. Although it is not the original method used, it is the one recognised and shown in woodcuts and engravings.

The charcoal burner made his preparations with great care, using knowledge passed down often over several generations. The site was chosen carefully and it was advantageous to re-use kiln sites because the ground beneath them became conditioned. The hearth or floor had to be level and if the kiln was made on a slope, the surface would be levelled. Advantage would be taken of any natural protection from prevailing strong winds. If these barriers did not exist windbreaks of latticed wood, bracken or sacking would have to be made.

The nature of the soil was important. A light and loamy soil would be sought and clay avoided. Clay is cold and will therefore lengthen the time needed for the burn, while a light, loamy soil is warm and will more readily absorb the moisture and tarry liquids given off by the wood during the process.

Uniform combustion depends to a small extent on air rising through the soil but animal burrows must be excluded. A shelter or hut was built close to the site with its door facing the kiln and the charcoal burners would live in this for the duration of the burn. The wood was cut into 3 - 4 foot (1 - 1.25 m) lengths and any over 6 inches (150 mm) thick were split. It was then stacked to dry for at least six months, since wet wood will not carbonise readily, and finally stacked in a wide circle around the kiln site. The stacks were roughly graded into small, medium and large dimensions to facilitate the uniform building of the kiln.

A heap was also made of fine branches, grass, straw, bracken or similar litter and a quantity of fine earth sieved and placed to one side. Fine earth and residues from previous burns would be ideal.

The central point of the kiln was marked with a peg and a string tied to this was used to mark out the circumference.

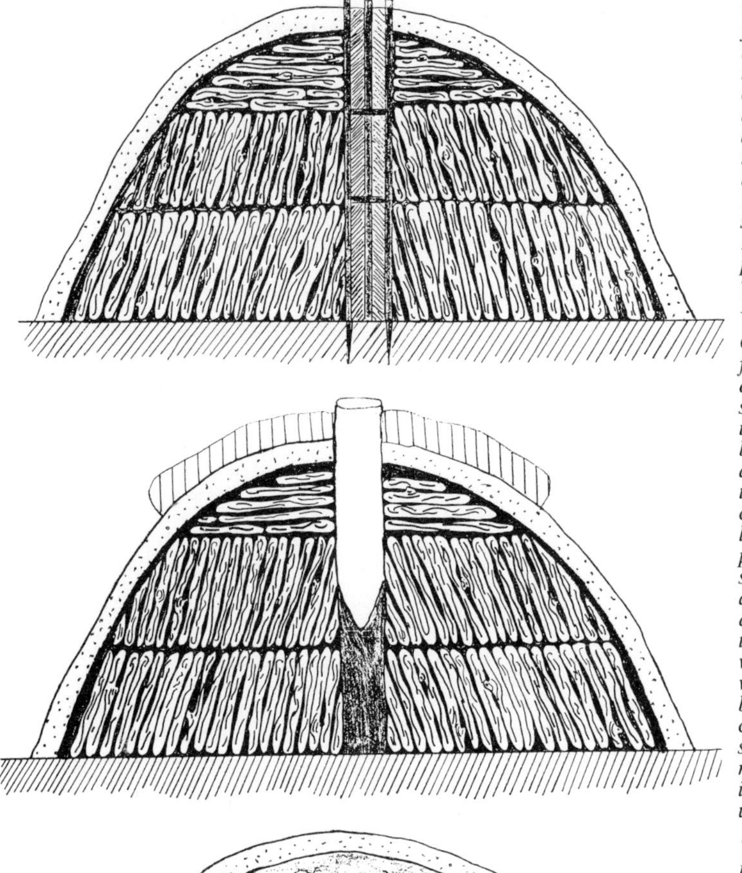

Cross-section of a forest kiln showing the three-pole chimney method. Three sharpened stakes were driven into the ground and held apart by rings of wicker or birch tied at intervals. This method was sometimes used in England but was more popular in Europe: it was seen in Italy in 1982.

Cross-section of a forest kiln showing the divided chimney peg system. This method involved building the base of the kiln around a thick pole driven into the ground. This was cut off after the first layer had been completed and another sharpened pole known as the motty-peg was driven into place. After the earth covering was complete this peg was drawn out and burning wood and charcoal poured in to start the burn. This method could be seen in the Lake District until the 1950s.

This cross-section illustrates the progress of the burn in a forest kiln. Spreading from the central chimney, the burn progresses in an enlarging cone shape until it reaches ground level. The kiln would contract steadily and evenly, the stack remaining stable since the unburned wood around the periphery would retain the shape.

This was necessary as a guide during building because the kiln had to be uniform and remain stable during the burn, when the wood would shrink and the heap subside.

The chimney could then be made. This could be in one of several forms and different areas had their own particular types: straight poles were often used, to be removed after building, or short pieces of wood laid horizontally in the form of a triangle, one triangle on another to a height of about 5 feet (1.5 m) to form a central flue.

The stack was then built around this flue, the construction being kept as round as possible. With tapering wood the narrow end would be placed on the ground, or the wood cut with a slanted end, so that less wood is in contact with the ground and the hot gases could circulate more freely. As it was built outwards the wood inclined slightly towards the chimney, with the largest pieces in the middle.

When the first layer was completed a second was built above it, with a greater inclination, and then a final layer to round off the top. The litter was then spread over the wood to bind the earth cover and prevent it falling into the kiln. The earth was spread evenly over this leaving a space around the bottom edge to admit air until the burn was established.

A fire of scrap wood and charcoal from a previous burn was made at one side. When this was reduced to a glowing mass, it was shovelled into buckets and poured down the kiln chimney. If a central pole had been used, this was first pulled out. Six or seven buckets might be needed. The chimney was left open for three to four hours to ensure that the kiln was well alight; if the wood was too green it would take longer. After a few hours damp patches would be seen on the earth covering, showing that combustion was under way and the remaining moisture being driven out of the wood.

The top of the chimney was then closed with litter and grass sods, a cap of these sods being formed over the top. The base was sealed and holes pierced in a circle below the cap. The fire having started in the central flue, it would now be drawn outwards and downwards by making holes lower down and sealing those above as the changing colour of the vapours indicated that carbonisation was progressing. At first the vapour would be white, turning darker and finally to a hazy blue as the wood was carbonised.

During the burn the kiln gradually subsided and the charcoal burners had to be ready to repair the earth surface where this cracked, or unwanted air would enter and spoil their work. By the end of the burn the kiln would have shrunk to about one-third of the original size. The experienced burner knew from the smell of smoke how the kiln was progressing; it was not wise to depend too much on the colour of the vapours.

When the burn was completed, the earth covering a small area of it was removed, two or three buckets of water were poured in and the earth was replaced. This was done in seven or eight sections and the steam generated helped to quench the charcoal. More earth was then put on and patted down, excluding all air. The kiln was allowed to cool for at least half a day or until it was completely cool before the charcoal was taken out.

When the kiln was opened, the charcoal was sieved to remove earth and dust and the unburned pieces or *brands* were removed. These brands were used to start the next kiln.

Variations on this method were used in different areas particularly when the by-products — the tarry liquors — were collected. Then the kiln would be built on a raised bed which was inclined towards a central gully, down which the condensed liquors would flow into a container. This was practised in some parts of Europe, notably Macedonia, and especially in Egypt and China: the Chinese tended also to use pit kilns with pipes to carry the liquors.

For production of charcoal rather than by-products hardwoods are preferred: European practice uses beech, oak, hornbeam and ash. Any wood can be used but these are preferred. For gunpowder alder, willow and alder buckthorn provide the best charcoal.

The charcoal burner's life revolved around the kiln. It had to be watched by day and night as any carelessness in maintaining the earth cover would cause the loss of some two weeks' work. The craft was often carried on by families, who were sometimes nomadic, and it was not unusual for one family to operate several sites with the women and children helping and carrying food to the burners.

Burning was mostly seasonal, particularly in the north of England, lasting from the beginning of April until November. In some places it was a summer and autumn occupation only. Wood was always cut in advance and stacked in *windrows* to dry, the charcoal burners turning to woodcutting during the winter when the sap content of the trees was at its lowest.

During the burning season the charcoal burners generally lived in a turf-covered hut on the site. This was usually a conical structure using poles 12 feet (3.66 m) or more in length. The diameter at the base was about 10 feet (3 m) and the poles were tied together at the apex. Girdles of wood woven or tied to the poles provided a base for sacking, over which was placed a layer of twigs to hold the turf covering, with the grass side uppermost. The turf at the base was built up in courses like brickwork and in some regions stone was also used. An opening was left for access, facing the kilns. Opposite this, at the 'rear' of the hut, in huts in the north of England there would be a fireplace. In the south a brazier was kept burning just outside the door. The doorway was usually closed by a curtain of sacks.

A bed or beds would be built on poles placed lengthways and fixed to stakes driven into the ground, the whole covered with springy brushwood and straw-filled sacks. On the more permanent sites the hut might have a table and outside there would be an oven consisting of a metal drum lagged with turf and with a metal plate acting as the door.

Although these huts were reasonably weatherproof and warm, wet weather did cause problems. One burner operating in Alderwasley, near Derby, between 1800 and 1820, brought up eight children in such a hut and in rainy periods would have to bail out the water which had collected during the night in the lower part of the floor area. It was said, however, that they preferred their warm hut to a cold stone-built cottage.

There were different levels of skill and commercial competence in this trade, as in any other. Some burners were nomadic, hiring themselves out to owners of woodlands or to fuel merchants. Others were more settled, operating in specific areas and hiring labourers for the season. These 'master' burners probably lived in more substantial cottages from which their wives and children would carry food to the burning sites. The wages paid are shown in records of estates and piecework was common. The Crown employed burners directly, records of the fourteenth century showing payment at the rate of threepence per man per day. In the years before the First World War a burner could earn between 20 and 30 shillings a week, just below the wages paid to smiths or other artisans. In 1750 in Nottinghamshire the rate of 3 shillings 'per dozen' — presumably a dozen standard-size sacks — was paid to jobbing burners.

A type of kiln used particularly in Austria in the eighteenth century and later in other areas. The wooden structure was coated internally with earth or a mixture of earth and charcoal dust. Short and misshapen pieces of wood which were not suitable for the normal forest kilns could be burnt in it.

A charcoal burner's hut of about 1890, believed to be in the New Forest. In the foreground, in front of the boy, is the typical woven oak basket (called a 'swill' in northern areas) used for all loading and carrying work on the site. This hut is very soundly built and appears to have a wooden door. This would be warmer than the caravan seen on the right. If this is the charcoal burner's family, they are dressed in their Sunday best clothes. Many families were raised in the woods in these conditions. Some burners were nomadic, hiring their services to landowners and farmers.

PROCESS DEVELOPMENT

All charcoal was originally made in the forest and carried from there to the place of use. This was because the charcoal made by those early methods represented around 15 per cent of the weight of the wood as cut from the tree while the volume was reduced by more than 25 per cent. It was therefore more economical to carry charcoal rather than wood.

The first method of production used was the pit kiln. Essentially simple in form, the surrounding earth retained the heat and a proportionally small top cover was needed. The site was carefully chosen: the soil neither too damp nor too dry, and well drained. There is little evidence of the size of these early pits or the degree of control exercised over the process, but since pit kilns still exist in several parts of the world the process can still be seen almost in its original form.

The pit was replaced by the earth-covered heap, or *meiler* as it is called in continental Europe. In Britain a woodland charcoal kiln site is still referred to as a *pitstead*. The covered heap method is more efficient: control is more exact and the process faster, and overall a better quality charcoal could be made. Moreover the by-product tars and liquors were more easily obtained.

The meiler quickly came to be a familiar sight as the demand for metal products grew. Charcoal burners were active in the Forest of Dean, the great forests of the Downs, the New Forest and elsewhere. The size of the earth kilns grew and it became necessary to improve their stability as the heap contracted during the process. This was achieved by building walls of turf around the base, added when the volume of steam reduced, or by encircling the base with wickerwork reinforced by posts.

Another form which was used later, particularly in Austria, was a rectangular wooden structure coated internally with earth mixed with charcoal dust. This reflected the early pit methods but also allowed the use of the short or misshapen wood pieces which would have made the normal heap unstable.

In some areas the number of burners working was considerable. In four of Edward I's demesne woods in the Forest of Dean in 1282 no less than nine hundred were operating. Five years before this the City of London had forbidden the entry of carts selling charcoal. Instead they had to offer their wares outside the gate at Smithfield or in a place provided at Cornhill.

Problems of pollution and sharp sales practice were recorded. In 1299 the 'unhealthiness of coal' was noted and in London's Guildhall in 1368 a statute was passed relating to charcoal weights. Under this statute, on conviction of selling sacks of 'less than one quarter in weight' the offender was to be 'put upon the pillory, and the sacks burnt beneath them'. Various individuals, many from Croydon and the Home Counties, were pilloried or placed in the stocks at Cornhill for these offences. It is not clear whether this statute was ever repealed.

During those years the techniques of charcoal manufacture varied very little but the volume produced continued to grow rapidly. Gunpowder became another market for the charcoal burners, requiring willow or alder charcoal. Gunpowder was used for the first time by English soldiers at the battle of Crecy in 1346. For the manufacture of gunpowder, the wood had to be cleaned of all bark and carefully checked by hand after carbonisation so that no grit was passed

In this photograph, from those believed to be from the New Forest, brushwood is being used as a windshield. The kiln has just been started up and the burner on the top is making his final check on the seal over the chimney before climbing down, removing his ladder and starting on the careful work of controlling the burn. His companion appears to be clearing the ground around the kiln, giving them a clean working space.

The process is almost finished and the heap has contracted. Note the shovel and rake held by the men, who are preparing to open the earth covering and rake out the charcoal. The wickerwork windbreaks can be clearly seen, kept in place to prevent gusts of wind from re-kindling the fire. The woman on the left appears to be assisting by removing wood from the kiln area, which has been carefully cleaned and swept clear of combustible debris. This cleanliness and care was typical of a good charcoal burner, who knew how quickly a ground fire could spread.

to the powder mills.

In 1414 Henry V ordered willow charcoal preparation for gunpowder but the use was still relatively small and the quality probably doubtful: several centuries were to pass before proper production control was established. Henry needed large quantities of hardwood charcoal for making armour and equipment and for his armaments. Every arrowhead was forged from iron and hardened, both processes necessitating the use of charcoal. It was this method that produced the arrows which proved so effective at Agincourt in 1415.

By the end of the fifteenth century ironmaking techniques had progressed to the point at which cannon, previously made by forging or strapping strips of iron into tubes, could be cast in one piece. The rapid destruction of the forests for shipbuilding, house building and charcoal making caused concern and charcoal burning came to be regarded as a public nuisance because of the great volumes of smoke produced. A law passed in 1558 prohibited the felling of trees to make charcoal for iron smelting, but by this time the supply of small wood from wood cutting operations and coppicing was not sufficient and it was largely ignored.

However, in the early part of the seventeenth century the shortage of charcoal became acute. Blast furnaces, developed some hundred and fifty years earlier, used large quantities of fuel and since around eighty-five of these were working in Britain there was a great demand for wood supplies. Forges also demanded substantial quantities and it is reported that in Sussex alone one hundred and forty forges each used over five tons of charcoal every week.

The process of charcoal making now began to undergo a period of change. Studies of the process and its by-products were being intensified. In 1658 came the publication of Glauber's *Miraculum Mundi* in which he identified the pyrolig-

A demonstration forest kiln on the Weald and Downland Open Air Museum site at Singleton, Sussex, in 1972. The burner has used straw to bind the earth covering.

neous acid produced in the charcoal process with the acid in vinegar, acetic acid. In 1661 Boyle described the separation of a 'spirituous liquid' from this pyroligneous acid: wood spirit.

These researches led to improvements in kiln design. The peak period of the European charcoal industry was in the seventeenth and eighteenth centuries but critical wood shortages and rising prices made new and more efficient methods necessary.

In 1709 Abraham Darby, working at Coalbrookdale, adapted a small charcoal blast furnace to use coke. He made this from mineral coal by kilning in open heaps after the manner of charcoal burners, referring to this as *charking*. The product was a relatively pure carbon; it was easy to produce and cheap coal was readily available. A patent on the coking process had been taken out by Dud Dudley in 1622 but never commercialised. The use of coal for iron smelting had been tried many times with poor results, mainly because of the sulphur content of the coal. It took the energetic Darby to develop the technique to the point of commercialisation but it was still many years before the new fuel was widely used, and even as late as 1788 only two thirds of the blast furnaces in Britain were using coke.

This development did not, therefore, have an immediate effect on the requirement for charcoal. It was balanced by the growing importance of wood charcoal by-products, which led to the manufacture beginning to be concentrated in larger units on an industrial basis. The process began to move out of the forests: where formerly the plant was taken to the wood, now the wood was being carried to the plant.

Interest in charcoal itself was also stimulated. Although its use for iron and steel production was diminishing, it was still necessary for hardening and alloying and in the production of non-ferrous and precious metals. Until 1770 it was the only satisfactory material for use in the conversion of cast iron into malleable iron. The 1797 edition of the *Encyclopaedia Britannica* described the manufacture and uses of charcoal and, in the description of its properties, argued that charcoal 'is indeed the phlogiston ... so long sought in vain'.

More pragmatic organisations were already putting research into the by-products to practical use. In Rotherham in 1796 the Shirley Aldred organisation was founded, producing charcoal and using the acid by-products to produce acetates of iron, lead, copper and calcium. The company is still in the wood distillation industry and is today the largest charcoal manufacturer in Britain.

The original earth heap methods continued in wide use, as they are today in less developed areas of the world. In the eighteenth century, modifications to these included the use of brick bases inclined towards the centre, with drains for the by-products. The drains led to external tar pits, closed to prevent air passing back into the heap. This system was popular in Poland, Russia and the Landes area of France, where pine wood tars were extracted. In 1801 Brune, a designer working on improvements in carbonisation methods, devised an extension to this by building an annular brick wall to replace the turf or wooden external supports. It was reported that the charring of the wood was expedited and the charcoal yield improved but the process was slower.

Variations on these systems were manifold, although the essential shapes of the circular and rectangular meilers were retained. Wooden structures with a variety of internal coatings and brick or stone walls with coated wooden covers began to be used on fixed sites. A simple method designed by Foucand around 1810 was the covering of the circular heap by internally coated wooden panels, tapered towards the top and fitted with a lid some 10 feet (3 m) in diameter. From this lid a pipe took the gases to a series of large wooden kegs in which the crude by-products were collected.

Domed brick kilns now appeared, de-

Starting the burn. Shovelfuls of burning wood and charcoal are being poured into the chimney aperture at this demonstration at the Weald and Downland Open Air Museum in 1971.

LEFT: *Cross-section of an early type of coal carbonisation kiln. The shape and earth covering of the stack followed the pattern of the charcoal burners but a brick chimney with regular apertures was used. Coke was made in this type of kiln until, as knowledge of the potential by-products was gained, metal-bound brick chambers and ultimately metal retorts were used.*

BELOW AND OPPOSITE, ABOVE: *Side and end views of a charcoal retort as used at the Royal Gunpowder Factory in the mid nineteenth century. The closed drums, known as slips, were made just over 3 feet (1 m) in length. The drums (A) were filled with selected wood, usually alder dogwood, and run up to the opening of the retort cylinder (B). Rollers on the carriages allowed the drums to be pushed easily into the retorts. These retorts were made of cast iron and fitted with a pipe (C) which led to a horizontal pipe behind the set of three cylinders, through which the gases were passed to a vertical pipe connected into the furnace (D). If the retorts were charged in proper rotation a supply of gas would always be available to supplement the furnace fuel. The space between the retort brickwork and the retaining wall (E) was filled with sand, which would retain the heat so that the tar-filled gases would not condense and quickly block the pipes. This sand was easily removed for cleaning and replacement of the pipework.*

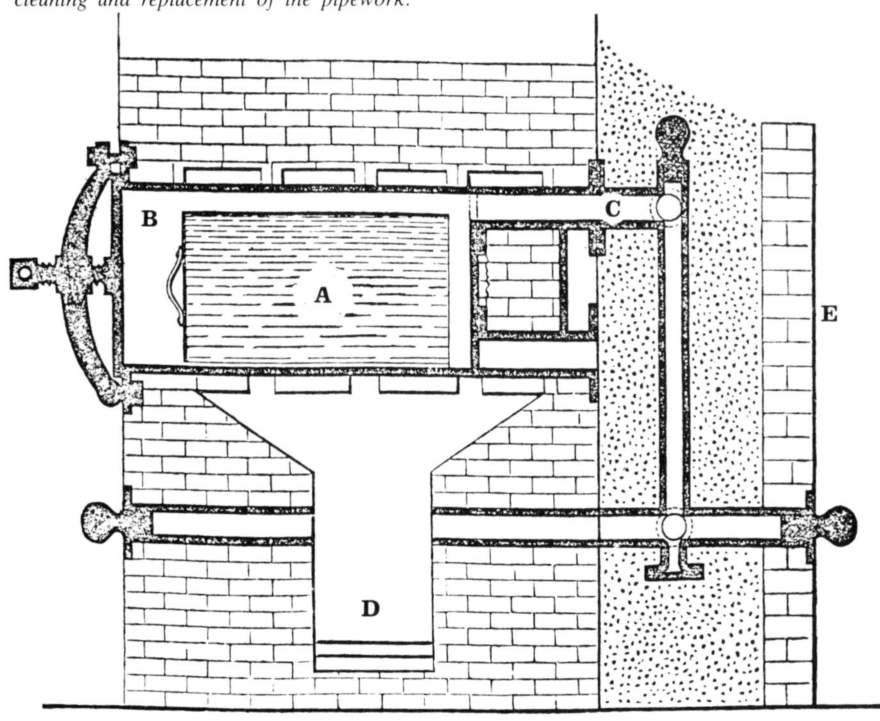

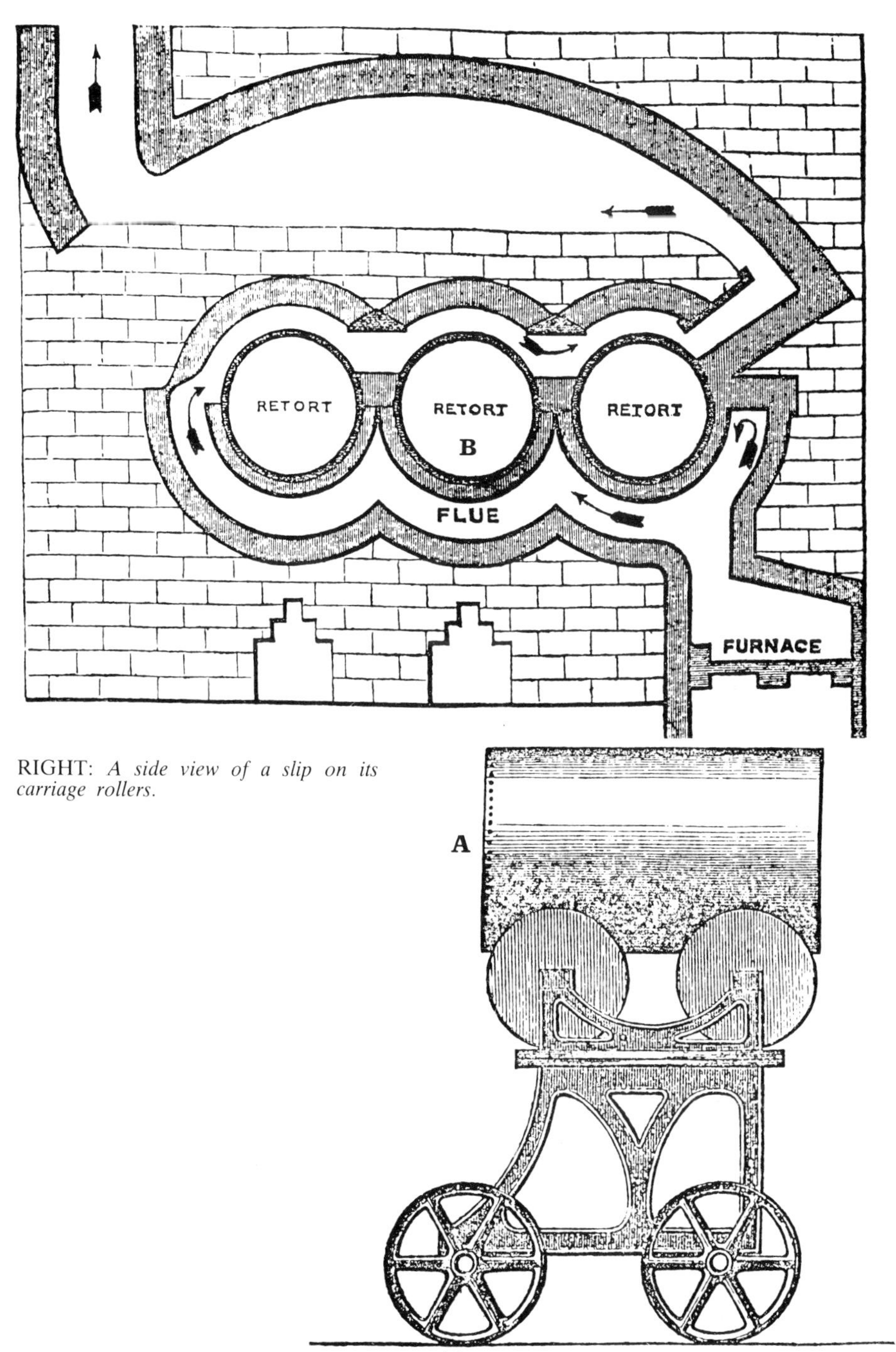

RIGHT: *A side view of a slip on its carriage rollers.*

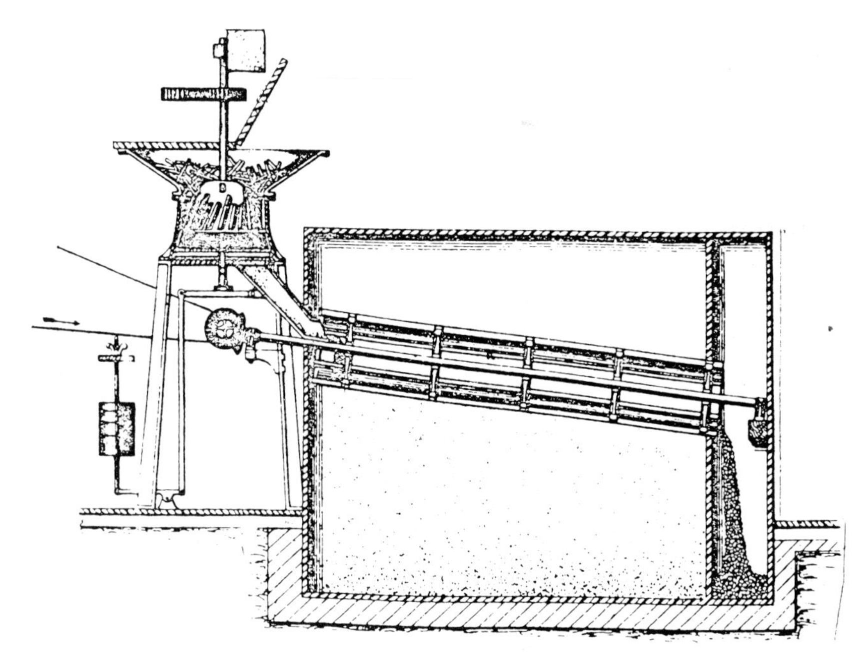

Charcoal crushing and sieving equipment, 1860-70. The charcoal would be reduced to granules and dust to facilitate the grinding process in gunpowder production.

veloped from the methods of Brune and others. External heating systems were used, with ports connecting the furnace to the kiln. Firing was by wood or coal. The first of these was constructed by Reinhold Freiherr von Reichenbach, who carried out work on this subject between 1820 and 1860 and published his findings on the components of wood tar in 1835. Brick kilns enjoyed only a brief period of importance, however, and were displaced rapidly after 1800 in Europe by iron retorts. In other areas they are still in use; Brazil has thousands of domed brick kilns and they are also still used in North America.

In 1810 Mollerat Frères developed a design based on a wrought iron cylindrical retort of 3 cubic metres (7.36 cubic feet) capacity, built into a brick chamber which allowed hot gases from the fireplace below to circulate around the cylinder before passing out through the chimney. It was loaded and discharged through a manhole in the lid. Gases from the interior were led out to a condenser. This was said by French writers to be the first plant invented for the extraction of the acid by-products from closed vessels, but perhaps they did not know of other systems in use elsewhere. This retort was modified by Kestner, who improved the throughput by use of a lid held by removable clamps and forming a door at the base for the discharge of charcoal, while still hot, into iron extinguishing boxes.

A portable retort system soon developed from the fixed retorts and was widely adapted. This was an ingenious method in which wood was loaded into a metal cylinder which was then sealed and lowered into a brick chamber incorporating a firebox. Distillation then proceeded, and on completion the cylinder was withdrawn by a travelling hoist sys-

A modern type of steel kiln being used as the basis of a new charcoal-making scheme in Egypt. This can utilise a wide range of wood shapes and produces better and cleaner charcoal than the local earth kilns. When the lower section has been filled (above), the upper section is lifted into place and filling continued to the top (below). Several such kilns are used on each site. Around 3 tonnes of wood are needed to fill the kiln and the process from filling with wood to removing the charcoal takes two days, a great advance on older methods.

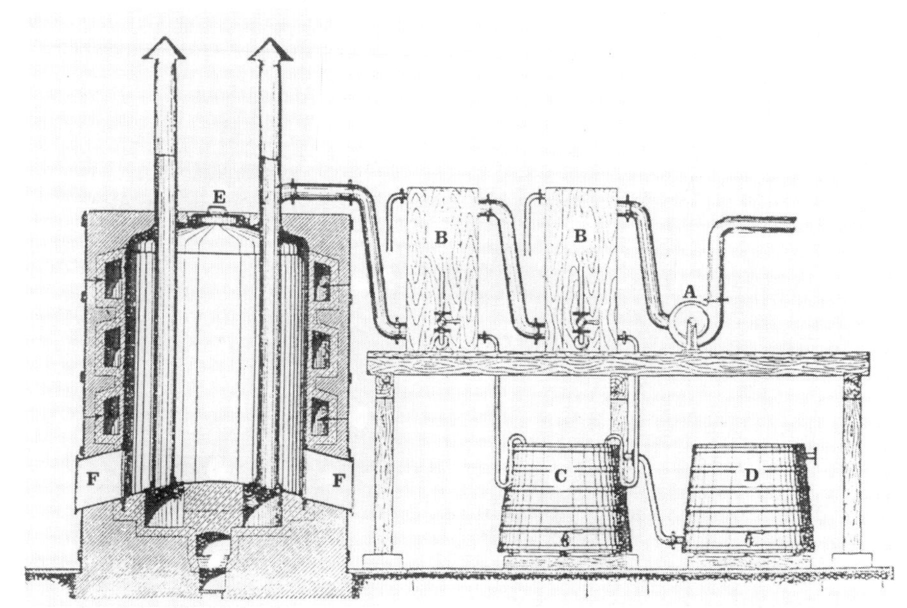

ABOVE: *The Bosnian Meiler oven. This unit had a capacity of 50 cubic metres (1765 cubic feet) of wood in each charge and the gases were drawn by a fan (A) through a crude condensing system (B). The tar and liquors ran by gravity into the first wooden vat (C), in which some of the wood tar settled out. The partly de-tarred pyroligneous acid ran from the top of the first vat into the second (D). The uncondensible gases (carbon dioxide, carbon monoxide, methane and some hydrogen) were taken through the fan and could be used for heating or lighting. Wood was introduced through the top opening of the kiln (E) and discharged at the base through the side openings (F).*

BELOW: *A horizontal retort. These were built in pairs with a common condensing system. The uncondensible gases were piped under the retorts and burned to reduce the amount of coal or other fuel needed for the process. These units were very laborious to charge and discharge.*

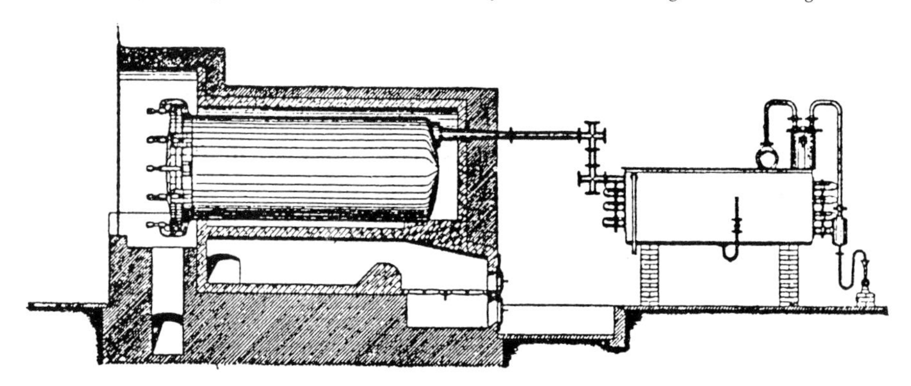

tem on an overhead gantry and replaced by a fresh cylinder of wood. The brick chamber remained hot and useful heat was therefore conserved. Since the metal cylinders heated up and cooled down quickly, the process was speeded up. A similar design was used by Turnbull shortly afterwards in Camlachie, Glasgow, and the system remained in use until the company closed in 1964-5. In 1849 Turnbull took this method to the United States, where he erected the first proper wood distillation plant in America at Milburn, New York. Plant of very similar design is used today in south-east Asia to make charcoal for the metals industries.

In 1819 Reichenbach designed what was said to be the first kiln in which the process heat was transmitted through metal walls. This could be disputed by reference to the Mollerat Frères design and the closed vessels or pipe kilns used thirty years before in Britain for production of gunpowder charcoal, following the suggestions of Richard Watson (1737-1816), Bishop of Llandaff. Lieutenant-General Congreve reported on the super-

A portable retort system, such as used by Turnbull and others. The retort chamber (A) was kept hot, the metal kiln bodies being charged with wood and discharged externally and lifted into and out of the chamber as required. After lifting the body into place, the gas pipes (B) had to be bolted to the top flange on the cover (C). This was a cumbersome arrangement but the system cost less than a fixed plant of similar output.

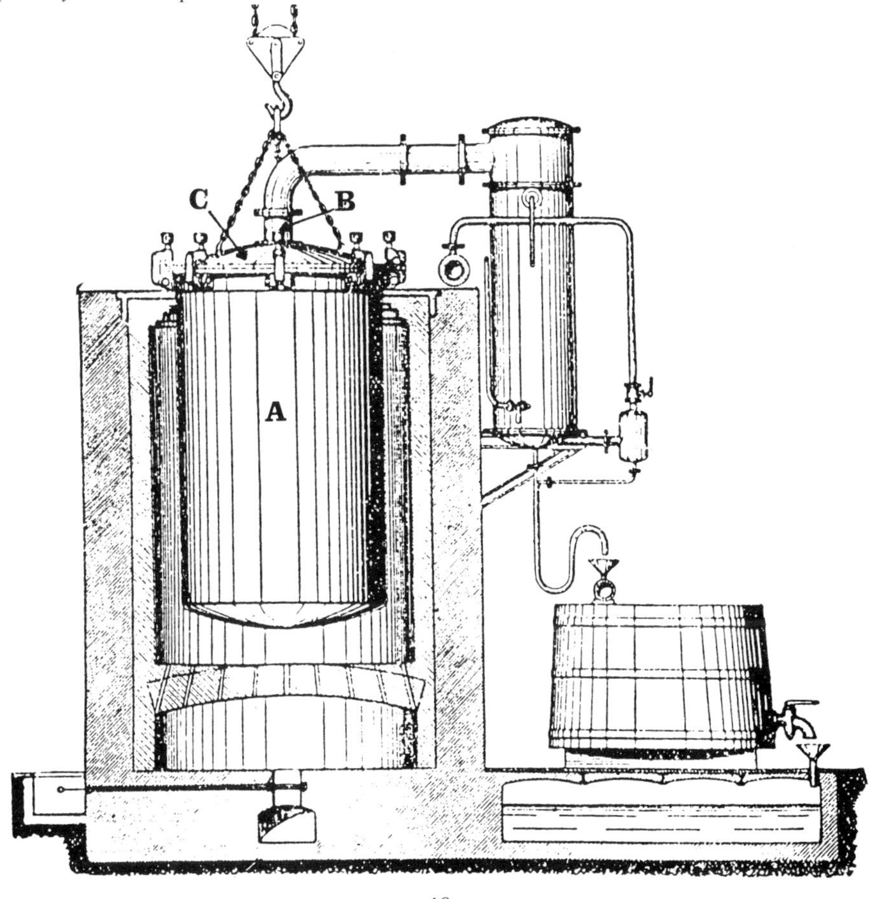

ABOVE: *Wood distillation works in the Forest of Dean, about 1920. The Meyer cylindrical kiln is in the masonry chamber seen under the roof of the building on the left. In the right foreground is the cooling chamber with its doors swung open and in the foreground is a wheeled cradle of charcoal waiting to be emptied. The level of the charcoal in the cradle can just be seen. When this was charged with wood it would be tightly packed up to the top. In the background is the by-product building which held plant for the manufacture of calcium acetate and distillation columns for the wood spirits.*

BELOW: *A semi-portable kiln of Belgian design. Its only advantages were simplicity of design and the ability to add extra sections if needed. The unit illustrated would hold 15 cubic metres (530 cubic feet) of wood, probably 8-9 tons. The date is unknown although probably 1920-30.*

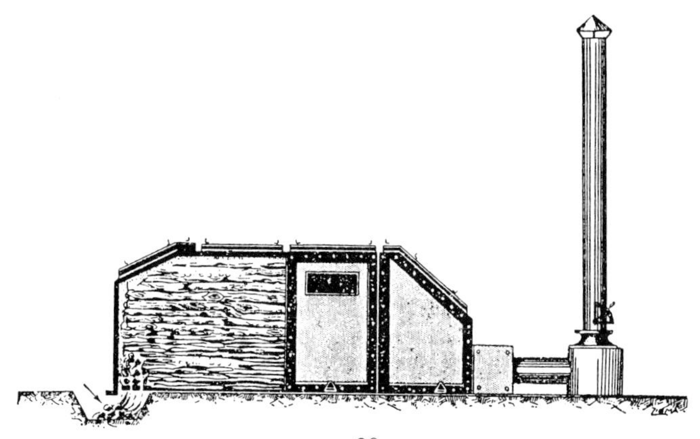

ABOVE: *A set of three tunnel retorts. The charging doors and outer insulating doors are lifted by wire cables to allow charging or removal of the wood cradles. The below-ground furnaces were stoked from the other end. The furnace chimney and the condensing system for the wood gas can be seen on the right. Condensed liquids were piped to wooden vats and left for twenty-four hours, during which time most of the tar would precipitate. By heating the liquor more tar could be removed and the wood spirits extracted. The de-tarred liquors were reacted with lime to make calcium acetate, which was then spread on heated concrete floors to dry.*

BELOW: *An automatic batch kiln with internal flues built into the lid. A Belgian design of the 1930s, this unit was designed to draw the heat from the centre to the exterior parts of the wood charge. It was not popular and did not give the same measure of control as the simpler units with exterior chimney systems. It was however much larger, holding 12 tonnes of wood. Many such designs were brought on to the market in attempts to offset the rising cost of labour, each claiming to need less attention during the process.*

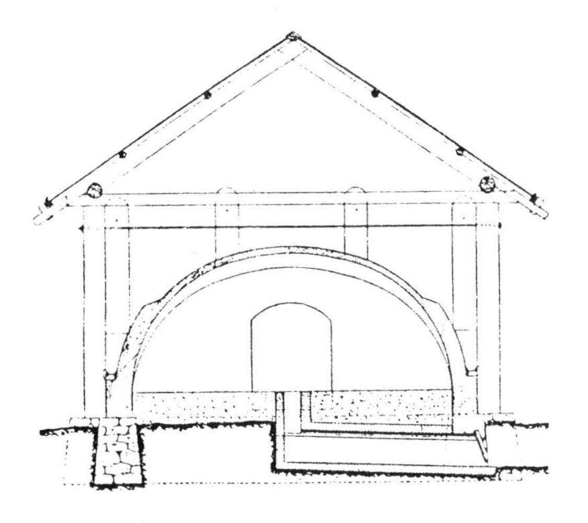

LEFT: *The Schwartz oven. This was in use in Scandinavia in the early twentieth century and has similarities to American designs. Two furnaces in the sides of the unit supplied hot gases which were led by a flue into the centre of the base. From this the gases circulated upwards through the wood.*

BELOW: *Three portable kilns are seen in this advertising illustration. The date is unknown but from the appearance is late 1920s or early 1930s. The Tattersall name persisted and cylindrical kilns of that name were still produced in 1970.*

iority of the charcoal made by this method. There are descriptions and illustrations of the cylinder retorts and liquor recovery system in use at Faversham in 1798.

Iron retorts designed by Hessel, who made notable contributions to the progress of charcoal development, were introduced in Germany and Russia around 1850. Similar retorts were in use in England and Austria. Both vertical and horizontal types were used, each with a single door. The horizontal types were simpler to discharge since the workmen did not have to climb down into them to remove the product and they could be rotated on the brick supports when the surface exposed to the external furnace became distorted or damaged by the flue gases. For the next thirty years there was little change in designs but great advances were made in the treatment and refining of by-products. Multiple retort systems mounted in brickwork, with a variety of flues and baffles to prevent direct impingement of firebox flames on the metal retorts, were noted in all major European countries.

ABOVE: *Charcoal burning on a fixed kiln site requires large quantities of wood. This is stacked to dry for periods ranging from three months to a year before use.*

RIGHT: *A Blair tunnel retort in the Forest of Dean. The outer and inner doors have been lifted and the hot wood gas has immediately burst into flames. The operator has attached the winch cable and chain for drawing out the cradles and will spray water on to these as they are pulled out of the retort and into the cooling chamber opposite. The retort will immediately be charged with another set of wood cradles and, since the unit remains hot, the distillation process will start quickly. Two sizes of these retorts were in use on this site. Of the five retorts, two took three cradles and the remaining three took five cradles per charge. The larger units were operated on a forty-eight hour cycle and the smaller could occasionally be cycled in twenty-four hours.*

ABOVE: *Blair, Campbell and McLean tunnel retorts at Worksop, Nottinghamshire, between 1940 and 1950. The insulated outer door, known as the storm door, is in the closed position to conserve process heat. When this is lifted the inner door of the retort proper can be unclamped and raised for the entry of the wood cradles. Each kiln took five of these cradles. When these are withdrawn at the end of the carbonisation cycle the charcoal contents will be below the level of the slatted sides. These are lifted off for removal of the product. These kilns were taken out of service in 1958 when a new vertical continuous kiln was started up on this site.*

LEFT: *Wood cradles being wheeled into a Blair, Campbell and McLean tunnel retort. The door seen above will be lowered and clamped into place and the furnace in the chamber below stoked up to start the charcoal process. Intense heat, acidic gases and stress caused by rapid cooling and water sprays were very hard on the cradles and maintenance of the framework and retaining metal slats had to be carried out continuously. The base of the retort is at ground level for ease of charging and the furnace is below ground. Flames from the furnace were not allowed to come into direct contact with the metal of the retort, the hot flue gases being channelled around it by flues built into the brickwork.*

In the last years of the nineteenth century new designs of tunnel kilns or retorts came into use. Notable amongst these were the retorts produced by F. H. Meyer of Hanover-Hainbolz, one of which was installed in the Forest of Dean in 1913 together with a distillation plant for the production of charcoal, wood tar, acetate of lime and wood spirit. The retorts were cylindrical and fitted with rails to carry wheeled cradles of wood. The retort doors were hung on wire cables and when the cradles were in place these doors would be lowered and clamped into place. The retorts were around 56 feet (17 m) in length and 7 feet (2.3 m) in diameter, supported along each side in brickwork. Heat was provided by coal fires in a furnace below ground level and the hot gases from this passed under an arch running the length of the retort and up through portholes placed at intervals on each side. On completion of the process the door was raised and the cradles drawn out into a cooling chamber of the same dimensions. Water was sprayed over the sealed cooling chamber and on to the cradles as they emerged from the retort. The retort, which remained hot, was immediately recharged with fresh cradles.

Another design, by Blair, Campbell and McLean Limited, of Glasgow, used tunnel kilns built in the approximate shape of railway carriages and operated in a very similar way to the Meyer kilns. These were suspended by steel straps from beams across the roof of a brick chamber and allowed free circulation of the hot flue gases around the unit. The original Meyer kilns in the Forest of Dean were replaced by Blair kilns, which were also installed in Warminster, Wiltshire, and Worksop, Nottinghamshire. Blair kilns were exported throughout the world, although the last units known to be in use were in the Forest of Dean. They were closed down and removed in 1970.

All charcoal plant up to the 1930s was batch production plant or, in as much as some used removable cradles and retained process heat, semi-continuous. Efficiency had been improved by burning part of the produced gases under the retorts and the direction of research turned towards the injection of burning gases directly into the wood charge.

Rectangular oven retorts. This is a diagrammatic illustration of the type of retort shown on pages 23 and 24. The bodies of the retorts resemble railway carriages, suspended by steel straps from girders set in the brickwork of the retort housing. Rails are fitted internally for the wood carriages. Since the retort bodies hang clear of the brickwork, the hot gases from the firehole below can pass freely around the unit. These were efficient units and modified designs are being made for possible use in the developing countries.

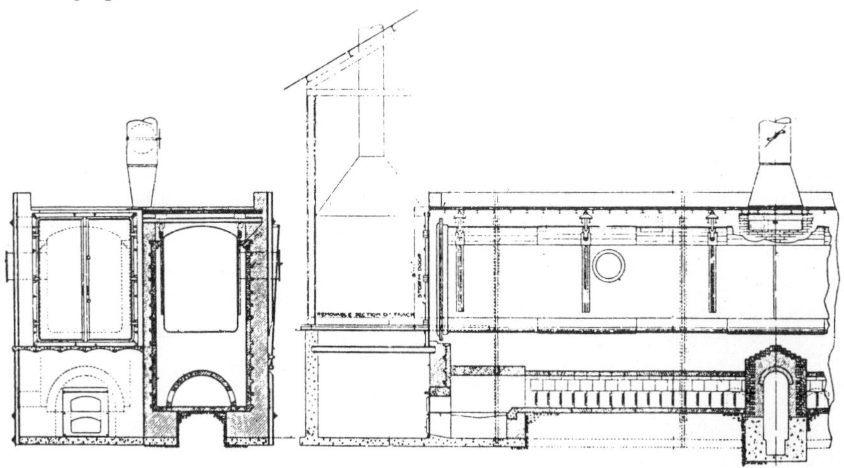

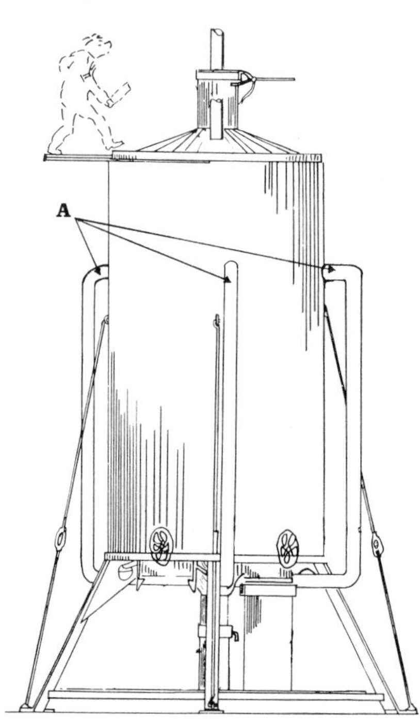

LEFT: *The Shirley Aldred retort under construction, 1967-8. The vertical retort can be seen in the centre of the building with the wood hoist housing on the right. In the lower centre are the three condensors for the wood gases.*

RIGHT: *The Cornell kiln, an American design for semi-continuous operation using exterior pipes (A) to direct wood gases from the upper to the lower parts of the kiln. It had fundamental problems and few were made. Increasing labour costs and a reluctance to handle dusty charcoal were the reasons for several such designs. The by-products were losing markets to the newer synthetic materials and the Cornell, amongst others, aimed to use the by-product gases by burning these directly in the kiln.*

During the 1920s and 1930s Reichert had developed a new system of internally heated batch retorts and these were eventually installed in Germany and Sweden. They were built in batteries, the gases from one being transferred to fire another. This was the first step towards an efficient continuous retort system, the search for which had been going on for some time. Success in this was achieved with the SIFIC process, developed by Lambiotte et Cie, who installed the first unit at Marbehan in the Ardennes in 1937. After the Second World War more of these were built in France and in 1958 one was started up in Britain at Worksop in Nottinghamshire. Apart from the original 1937 retort, all are still in use. They are internally heated vertical continuous units using the wood gas for heating and coupled to by-product recovery systems.

The most recent developments were made in the 1970s when the Lambiotte company produced a less costly vertical unit with automatic control, incorporating an integrated circuit control system. The Shirley Aldred company developed a horizontal continuous unit for processing previously unused small wastes — sawdust, nut shells, coffee husks and similar agricultural wastes. Both companies have installed these systems in other parts of the world with the object of using alternative raw materials to reduce the consumption of wood from depleted forests.

The Shirley Aldred and Company Ltd charcoal works at Worksop, 1958. The new vertical retort housed in the building on the right used the SIFIC process developed by Lambiotte et Cie in Belgium. In the centre wood is being taken through drying tunnels to reduce the moisture content further before being taken by internal hoist to the top of the retort. On the left the tall building contains the azeotropic distillation system which produced wood oils, naphtha, methyl alcohol and acetic acid.

USES OF CHARCOAL

In Brazil, south east Asia and areas without large supplies of mineral coal available, charcoal is the predominant fuel used in the smelting of iron and refining of steel. The ferrous metal industries of western countries use coke for smelting and refining but charcoal is still used in the manufacture of special alloys and carburising or hardening processes.

Its widest industrial use is in copper refining and in the domestic field it is cookery. It is the urban domestic fuel in many countries where wood is not suitable in small domestic stoves because of its bulk, low heat emission and smoke nuisance. In recent years in Europe the growing popularity of barbecue cooking has brought charcoal back into the public view, now packed in small and colourful paper bags and sold from supermarket shelves.

The porous qualities of charcoal were known to the ancient Egyptians and the Romans. Wood charcoal continued to be used until fairly recent times in the filtration of drinking water and at present a refined form of charcoal, known as activated carbon, is used.

Activated carbon can be produced from wood charcoal by two methods, known as steam activation and chemical activation. These enhance the porosity of the material by creating pores and capillaries of molecular dimensions. The resulting material has a very high internal surface and is capable of removing contaminants or colours from liquids and purifying gases. Materials other than wood are also used for producing activated carbon, notably coconut shell, coal and lignite. Some grades of activated carbon are found in gas respirators, cooker hoods and used for the removal of odours from air.

Charcoal still has some medicinal applications. Being porous, sterile and chemically inert, it was at one time used in dressings for wounds and suppurating lesions. It was also taken internally as a remedy for stomach and intestinal dis-

Bodmin Moor, Cornwall, 1982. Two portable kilns beside a forest road producing charcoal from forest wastes. The nearer of the two kilns has completed the burn cycle and is cooling down. The far kiln is halfway through the burn cycle and the vapour from the chimneys is thinning. These kilns each hold around 3 tons of wood and produce half a ton of charcoal in each forty-eight hour cycle.

orders. Diabetic biscuits, known as charcoal biscuits, are still used.

Bone charcoal is preferred for the decolourisation of sugar liquors in the manufacture of white sugar. Wood charcoal in granular and powder form is used in agriculture and horticulture: it is an ingredient in bulb fibres, lawn dressings, dog biscuits and animal feeds.

Particular types of charcoal, alder, willow and alder buckthorn are used for gunpowder and fuse powders. All the bark is removed before kilning and the charcoal is very carefully examined to ensure that no grit, stone or metal particles, which might cause a spark, are introduced into the powder mills. Willow sticks, the bark being peeled off immediately after cutting, are used to make artists' charcoal.

A horizontal continuous kiln. This was developed to make use of small waste materials such as nut shells, coffee husks, sawdust and chippings. The gases produced in the process are piped directly to the furnace and the heat from this is circulated around the kiln body.

Alternative materials. The briquettes shown here are made from charcoal produced from coffee bean husks in Kenya. These briquettes are in every way as efficient as wood charcoal briquettes and illustrate the potential use of agricultural waste to preserve forests.

CHARCOAL BY-PRODUCTS

The use of pyroligneous acid from charcoal kilns by the Egyptians and of wood tar by Roman shipwrights has already been mentioned. Wood tar and specifically pine wood tar or Stockholm tar feature in lists of naval stores from the earliest written records. Pine tar was identified in the remains of Henry VIII's flagship, the *Mary Rose,* raised from the bottom of the Solent in 1982.

Pitch for use by shipwrights was made by boiling the tar until a cooled sample showed the correct consistency. If it was boiled for too long it could produce a brittle pitch and under some circumstances would foam and harden into a coke which was useless for shipboard use. A powdered resin would often be added. This not only speeded up the process but made a better and longer lasting dressing. Shipwrights and ships' carpenters working away from their yards or with no suitable equipment would heat the tar in an iron pot, ignite it and when the tar had thickened sufficiently, extinguish it by throwing a cover over the pot.

By condensing the vapours from charcoal kilns, a wide range of materials can be produced. The liquid condensate is an aqueous mixture of pyroligneous acid and wood tar. When run into tanks and allowed to settle part of the wood tar separates, particularly if the liquid is heated. The remaining liquor can be poured off and distilled to produce pyroligneous acid, wood oils, wood tar and spirits. The wood oils are used in preservatives, solvents and some insecticides.

By distillation, crude forms of methanol or methyl alcohol are produced which were used as a denaturant for ethyl alcohol, as a solvent in the aniline dye industry or for the production of formaldehyde. The remaining pyroligneous acid can be treated to remove water, and the resulting acetic acid (formerly known as wood vinegar) has a wide variety of applications including the manufacture of cellulose acetate, methyl acetate, sodium acetate and lead acetate, and for dyeing and other textile processes.

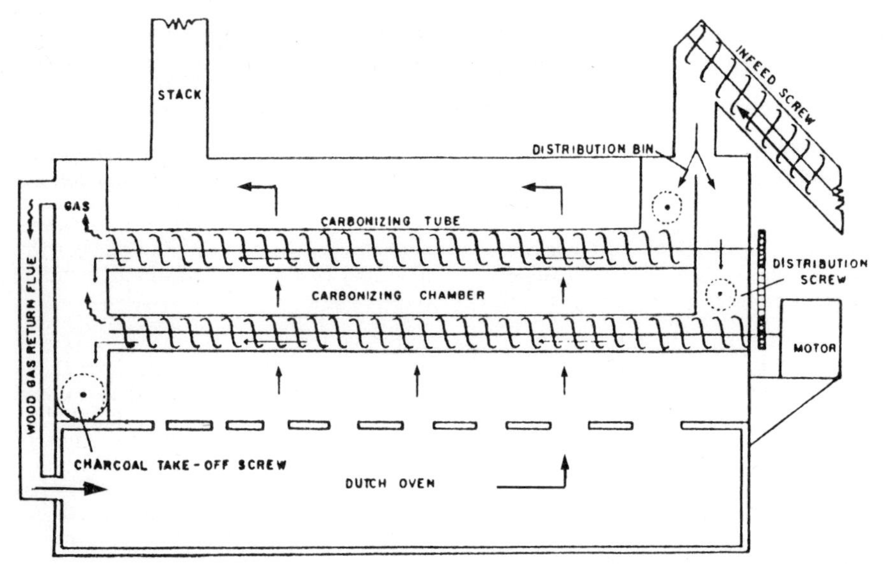

Flow chart of a continuous carbonising unit for sawdust, bark chips and shavings, developed by Svend Thomsen, a Danish-Canadian inventor and brought to commercial use in the 1950s. Screw feeders pass the material slowly through the tubes and the wood gas produced by the process is burned in the Dutch oven below to maintain the process temperature, which is held around 450 Celsius (850 Fahrenheit). The time required for the sawdust to be converted is one to two hours.

Stockholm tar and liquors were used in medicine and still have a place in veterinary work. In the production of charcoal by kiln methods, the recovery of these is simple but the other by-products were lost until seventeenth-century chemists began to take a more scientific approach. As has been mentioned above, Glauber identified the pyroligneous acid liquor with acetic acid in 1658. Boyle in his *Sceptical Chemist* (1661) went further and described the separation of a spirituous liquid from pyroligneous acid. This was wood spirit or methyl alcohol, the identification of which was finally confirmed by Taylor in 1812.

Following the work of Darby and others, William Murdoch introduced coal gas in 1792. In a sense he was instrumental in developing the wood distallation industry for it promptly adapted his techniques for the separation and collection of wood carbonisation by-products. Lowitz in 1793 was reported to have developed methods of purifying *wood vinegar,* presumably having found ways to remove the formic, butyric and propionic acid fractions present in the crude acid.

In Britain the partnership of John Aldred and Dr Thomas Warwick started work in Rotherham, developing and adapting systems for the production of chemicals using the wood distillates. They advertised their 'sugar of lead' (lead acetate), introduced the Turkey red dye and discovered a process for producing a fast yellow dye, reputed to be the only one known at that time. They also produced pyroligneous acid for the white lead industry, whose manufacturers employed it for the oxidation of blue lead.

In 1799 Phillipe Lebon patented a process of recovery of pyroligneous acid and wood spirit and devised the 'Thermo-lamp' which would use wood gas. The pace of development accelerated with the growth of the textile industry, which used the refined by-products in the dyeing processes, and the rapidly expanding chemical industries. The components of wood tar were ascertained by Reichenbach around 1835. In 1860 von Berg published his *Verkohlen des Hölzes* on the production of by-products from Meiler-type kilns.

By 1870 a chemically pure acetic acid was being obtained at a reasonable price from calcium acetate, itself produced from combining lime with pyroligneous acid. This same calcium acetate was being used ten years later in destructive distillation to produce acetone on an industrial scale.

Turnbull's plant in Glasgow had a demountable pipe system for taking wood gas from his metal retorts through a condensing system and using the condensed liquors to manufacture iron pyrolignite, used as a mordant in textile dyeing. This method was cheap and successful and was not abandoned until 1963-4.

By 1950 four or five main separation and refining systems were in use. Azeotropic distillation to produce wood oils, naphtha, methyl alcohol and acetic acid competed successfully with older methods, and systems for vacuum distillation and solvent extraction were perfected. The industry then suffered a heavy blow with the introduction of cheap and pure acetic acid made as a by-product by oil refineries. For some years the production of acid had been the mainstay of the wood distillation industry, its other products having been superseded by oil fractions or process changes. This final blow reduced the industry to a fraction of its former size. However, in the 1980s there are signs of activity and research into new methods, prompted by oil shortages and the carbonisation of alternative materials. The realisation that wood is a renewable source of energy has prompted government-sponsored research projects in western countries.

A tar distillation unit designed by the German company Meyer about 1920. The still body (A) is of cast iron and is heated directly by a fire (B). A large diameter pipe (C) delivers the hot residual pitch to a tank (D). From the dome or head another pipe leads the distilled vapours into a condenser (E) and the condensed liquors run down into the separator (F) which separates the water and the wood oils.

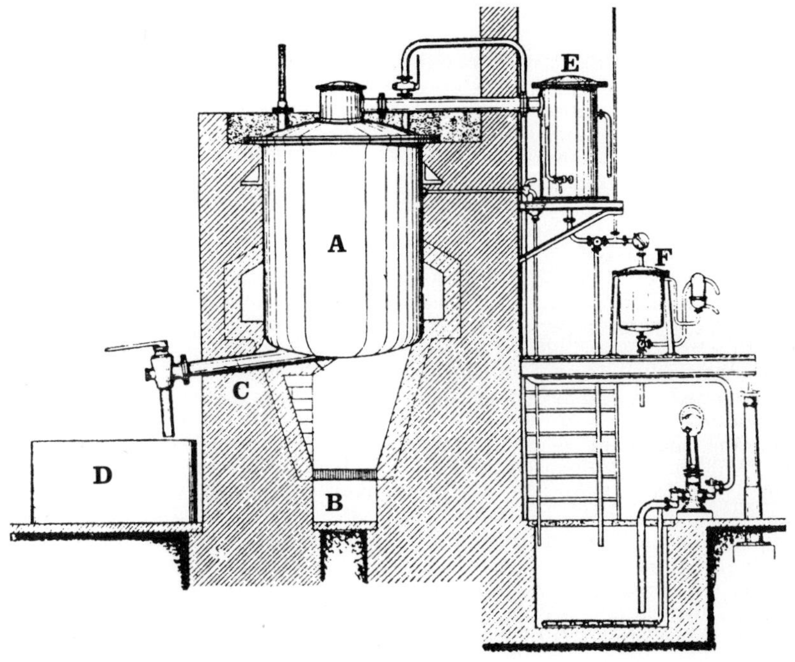

FURTHER READING

Agricola, G. *De Re Metallica.* Translated by H. C. and L. H. Hoover, 1950.
Armstrong, L. *Woodcolliers and Charcoal Burning.* Coach Publishing and The Weald and Downland Open Air Museum, 1978.
Bird, J. *The Medicinal and Economic Properties of Vegetable Charcoal.* John Churchill, 1855.
Bunbury, H. M. *The Destructive Distillation of Wood.* Benn Brothers, 1923.
Dickinson, H. W., and Straker, E. *Charcoal and Pyroligneous Acid Making in Sussex.* Paper read at Imperial College, 1937.
Klar, M. *The Technology of Wood Distillation.* (1903). Translated by Rule. Chapman and Hall, 1925.
Du Monceau, Duhamel. *Art du Charbonnier.* Circa 1760.
Scrivenor, H. *A Comprehensive History of the Iron Trade.* 1841.
Smith, D. M. *The Industrial Archaeology of the East Midlands.* David and Charles: Dawlish Macdonald, 1965.
Strand Magazine, 1895, Volume IX.
Tylecote, R.F. *A History of Metallurgy.* The Metals Society, 1976.

PLACES TO VISIT

Bewdley Museum, The Shambles, Load Street, Bewdley, Worcestershire. Telephone: Bewdley (0299) 403573.
Ironbridge Gorge Museum, Ironbridge, Telford, Shropshire TF8 7AW. Telephone: Ironbridge (095 245) 3522.
Kelham Island Industrial Museum, Alma Street, Sheffield, South Yorkshire. Telephone: Sheffield (0742) 22106.
Queen Elizabeth Country Park and Butser Ancient Farm, Gravel Hill, Horndean, near Portsmouth, Hampshire PO8 0QE. Telephone: Horndean (0705) 595040.
Weald and Downland Open Air Museum, Singleton, Chichester, West Sussex. Telephone: Singleton (024 363) 348.

A wood gas producer. Made by the English company Crossley in the 1915-20 period it operated on the suction principle, the quantity of gas generated being governed by the power required from the engine to which it was connected. Six units of this type, producing 190 brake horsepower each, were used in the Bristol Aeroplane works to drive six Crossley twin gas engines. This technology is also being updated and modern gas producer plants will operate with a wide range of vegetable and wood wastes.